Kinetic energy is the energy of motion. Anything that is moving has kinetic energy.

Potential energy is energy that an object has because of its position. It is energy that has not been used yet. The toy inside the box has potential energy to pop out.

READING FOCUS SKILL
COMPARE AND CONTRAST

When you compare and contrast you tell how things are alike and different.

Look for ways to compare and contrast forms of energy.

Some Sources of Energy

Energy is what makes it possible for things to move or change. Energy has many sources.

A moving car gets energy from gasoline, a fuel. You get energy from food. Food gives you energy to move, grow, and change.

Energy from moving water turns the wheel.

Lightning is electric energy.

Energy

Lesson 1
What Is Energy?. 2

Lesson 2
How Can Energy Be Used? 10

Lesson 3
Why Is Energy Important? 16

Orlando   Austin   New York   San Diego   Toronto   London

Visit *The Learning Site!*
www.harcourtschool.com

Lesson 1

What Is Energy?

VOCABULARY
energy
kinetic energy
potential energy

Energy is what makes it possible for things to move or change. The energy of moving water can turn a water wheel.

Plants get energy to grow from sunlight. Most of the energy on Earth comes from the sun's heat and light. The pictures show other sources of energy.

Tell three sources of energy.

Sound energy makes this alarm clock ring.

Electric energy lights up the sign.

Wind energy moves this windsurfer.

Forms of Energy

Energy can be grouped in two ways. It can be potential energy or kinetic energy.

Anything that is moving has **kinetic energy**, or energy of motion. When you move down a slide, you have kinetic energy.

Now think about sitting at the top of a slide. You are not moving, but you have energy. You have **potential energy**, or energy due to the position of an object. Your position gives you potential energy to slide down to the bottom.

Potential energy
The toy inside the box has potential energy to pop out of the box.

Kinetic energy
The toy has kinetic energy as it pops out of the box.

A battery has both potential energy and kinetic energy. But not at the same time. It has potential energy when it is not in use. It has kinetic energy when it is making something work.

Tell the difference between kinetic energy and potential energy.

Battery

Energy Changes

Think about a ball on a shelf. The ball has potential energy because of its position. Suppose the ball falls off the shelf and rolls across the room. Now it is moving. So the ball has kinetic energy. Potential energy was changed to kinetic energy.

Children have kinetic energy when they are jumping. ▶

Next, suppose you pick up the ball and bounce it a few times. Then you put it back on the shelf. The ball had kinetic energy when it was bouncing. But it has potential energy again when it is back on the shelf. Kinetic energy was changed to potential energy.

 How is the energy of a ball on a shelf different from the energy of a bouncing ball?

▼ A moving ball has kinetic energy.

Review

Complete the compare and contrast statements.

1. People get energy from _____ , and plants get energy from _____ .

2. A moving object has _____ energy, and an object that is not moving has _____ energy.

3. A _____ can have potential energy or kinetic energy depending on whether it is bouncing or sitting on a shelf.

Lesson 2

VOCABULARY
combustion
temperature

How Can Energy Be Used?

Combustion is another word for "burning." Some machines get energy from combustion. They may burn gas, wood, or coal to get energy.

10

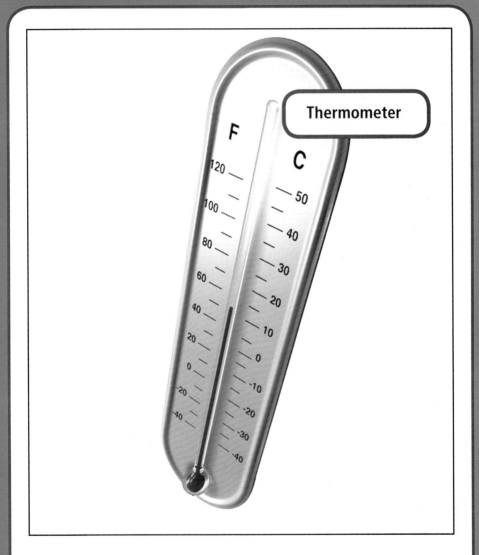

Thermometer

Temperature is the measure of how hot or cold something is. A thermometer measures temperature.

READING FOCUS SKILL
MAIN IDEA AND DETAILS

The main idea is what the text is mostly about. Details tell more about the main idea.

Look for details about the ways energy is used.

Using Energy

Energy makes things move or change. Plants use the sun's energy to grow. You get energy from the plants when you eat them. The energy moves through your body. This energy keeps you healthy. It helps you grow and move. You cannot live without energy.

▼ You get energy from eating plants.

Machines need energy to work. One way machines get energy is from **combustion**, or burning. A car engine burns gasoline for energy to move. A stove can burn gas for energy to cook food. A fireplace burns wood for energy to heat a room.

 Tell what happens during combustion.

▼ This train burns coal to help it move.

Measuring Energy

There are many ways to measure energy. How you measure energy depends on the type of energy.

A thermometer is used to measure temperature. **Temperature** is the measure of how hot or cold something is.

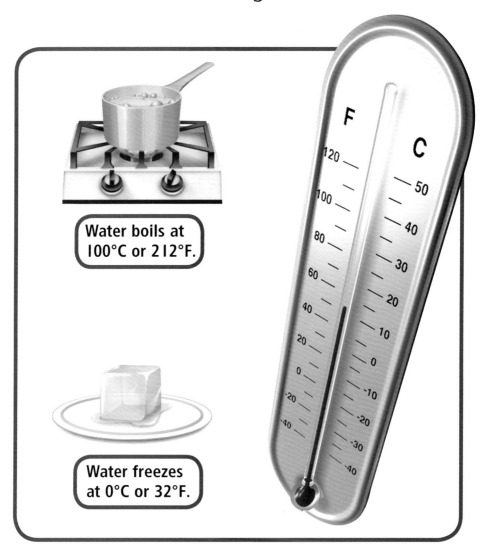

Water boils at 100°C or 212°F.

Water freezes at 0°C or 32°F.

A thermometer can only measure temperature. The tools on this page measure other types of energy. An anemometer measures wind speed. A light meter measures the strength of light.

Anemometer

Light meter

Focus Skill **What are some kinds of energy that can be measured?**

Review

Focus Skill **Complete the main idea statement.**

1. _____ makes things move or change.

Complete the detail statements.

2. Plants use energy to _____.

3. People get energy from eating _____.

4. Machines need _____ to work.

5. Different kinds of tools can _____ energy.

Lesson 3

Why Is Energy Important?

VOCABULARY
resource
fossil fuel
nonrenewable resource
renewable resource

A **resource** is something found in nature that people can use. Charcoal, made from wood, is a resource used for energy.

A **fossil fuel** is a resource that comes from the remains of plants and animals that lived long ago. Oil is a fossil fuel.

A **nonrenewable resource** is a resource that cannot be replaced during a human's lifetime. Coal is a nonrenewable resource.

A **renewable resource** is a resource that can be replaced. Energy from the sun is a renewable resource.

READING FOCUS SKILL
MAIN IDEA AND DETAILS

The main idea is what the text is mostly about. Details tell more about the main idea. Look for details about saving energy.

The Importance of Energy

All living things need energy. You are growing taller and bigger. Growing takes energy. You use energy when you walk, run, and play, too. You use energy to breathe. You even use energy when you sleep.

People get energy from food. ▼

A community needs energy, too. It takes energy to cook food. It takes energy to light, heat, and cool homes, schools, and office buildings. It takes energy to run cars, trucks, buses, and trains.

 What are some ways that communities use energy?

▼ Communities use energy at night, too.

Ways to Save Energy

A **resource** is something in nature that living things can use. Some resources are used to produce energy. Coal, oil, and natural gas are energy resources. They are called **fossil fuels**. A fossil fuel is a resource that comes from the remains of plants and animals that lived long ago. Fossil fuels are used to produce energy.

Oil pump ▼

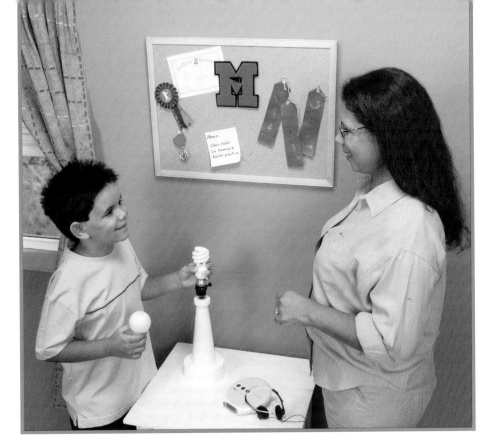

▲ This boy saves energy by replacing old light bulbs with ones that use less energy.

Once fossil fuels are used up, there will be no more of them. Fossil fuels cannot be replaced during a lifetime. So it is important to save them to make them last longer. You save resources when you use less energy.

 Tell why it is important to save fossil fuels.

Other Energy Resources

It took millions of years for fossil fuels to form. Once fossil fuels are used up, more cannot be made. They cannot be replaced. Resources that cannot be replaced are called **nonrenewable resources**.

▼ Charcoal is a nonrenewable resource. People burn it to cook food

Some resources can be replaced. They are called **renewable resources**. Wind is a renewable resource. Wind turns windmills, which produce electricity. Solar energy, or the sun's energy, is also a renewable resource. It is used to produce electricity.

▲ Windmills

 Name some renewable resources.

Review

Complete the main idea statement.

1. Energy _____ are important.

Complete the detail statements.

2. Coal, _____ , and natural gas are energy resources called fossil fuels.

3. _____ resources cannot be replaced.

4. Wind and solar power are _____ resources.

GLOSSARY

combustion (kuhm•BUS•chuhn) Another word for burning (13)

energy (EN•er•jee) The ability to make something move or change (4)

fossil fuel (FAHS•uhl FYOO•uhl) A resource that comes from the remains of plants and animals that lived long ago (20)

kinetic energy (kih•NET•ik EN•er•jee) The energy of motion (6)

nonrenewable resource (nahn•rih•NOO•uh•buhl REE•sawrs) A resource that cannot be replaced during a lifetime (22)

potential energy (poh•TEN•shuhl EN•er•jee) Energy that has not been used yet (6)

renewable resource (rih•NOO•uh•buhl REE•sawrs) A resource that can be replaced quickly (23)

resource (REE•sawrs) A material that is found in nature and that is used by living things (20)

temperature (TEM•per•uh•cher) The measure of how hot or cold something is (14)